Aliens UFOs and other life Forms

The question the future of humans

By

UFOs

First to answer the question do they exist? And if they do then it must go hand in hand that so do aliens.

Answer is 'yes' they do exist

Over hundreds of years there have been sightings of UFOs and indeed many pictures of such. Many different shapes and sizes and all manner of colours. So really nothing strange there and if you look at the many flying machines we have here on earth there are just as many shapes. So too aliens as after all the UFO needs someone to fly it!

As above there are those who clime to have seen an alien. And again pictures showing many forms and shapes. Why they should look so different from us is simple. There form is in tandem with the environment in which they live. For example a fish looks like it dose because of where it lives and how it lives. If you saw a fish walking down the street you would think 'what is that' (if you had not seen a fish before) There body has to accommodate organs that can function in order to support life within the environment where ever they live. So this explains there different forms. I can imagine when they look at us they think 'strange looking creatures these aliens' and really if you take a look at us humans we are all very different.

Ok! So they exist how do they get here?

To understand how they get here we first have to stop thinking we live in a 3 dimensional world. I believe we are living in a world of many dimensions. To say how many I think is madness as I believe that dimensions are created all the time and that we are living side by side in parallel worlds. I also think that 'black holes' play a key role here.

You see if a UFO has its home 4 million light years away then how can it pop over to earth and home again? I believe black holes are somehow gate ways through time zones and there are many gates not just black holes. It is really quite easy to understand that we are indeed living side by side to many other life forms. These life forms have been around longer than we have and there for have far more intelligent than us humble humans.

I also believe that much that we think we know is in fact very wrong. As many times our science has got it wrong and we have to re-write it. We do not yet have the technology and knowhow to go where we need to be to prove anything beyond the moon and our telescopes don't give us enough info as yet. But one day who knows? It's also interesting to note that when UFOs are seen many say that they just disappear like magic the reason is clear they simply go through a time gate and I think they are always near a gate as I said there are many dimensions.

What do aliens want from us?

In truth nothing! They are just looking with interest at other life forms. The sad thing is if we ever get to where they live what the hell will we do? If aliens wanted anything from us they would have made some sort of contact long ago. In truth I think they just want to keep an eye on us. If you look at the evidence that they exist then you have to believe it. But as for these UFO hunters my advice is 'give it up' I do not believe any government is in contact with other worlds any more than you or I.

People who say they were abducted

Oh I love this one! The problem with people is they want to be in the lime light 'sorry' but again why would aliens want to abduct anyone? Ok I hear you shout 'to experiment on us. Well with the technology they have I am sure a simple x-ray or scan will do the job without the need to cause too much excitement. I can not in my wildest dreams imagine aliens abducting a person or anything. Another thing is why do they all say 'I was on a white table and they were looking at me ext ext' you see it all sounds like a scene from a film. It would be a bit more believable if they said something like ' I was shown around that spaceship and given food. They spoke to me using pictures and sounds. The food was strange but very nice. But no! We get 'I was on a white table bla bla bla. So really my take on this is that these people have vivid imaginations or just want to be in the light or make some cash. There was even one guy, who said the aliens had left a chip or something in his arm, turn out to be something he had put there. So how can you believe such crap?

Are aliens friendly?

I think if they ever made contact with us we need to be ready. I do not think they would come in peace but who knows. I do think they would try to communicate with us somehow and it may work but I feel that if they come here then they will want more than a cup of coffee. I maybe wrong but this is my opinion and I think many think the same. This is why I say they have never abducted anyone because that would do damage to them like the secrets out!

Other life forms

We can say 100% that there are other life forms out there or next to us in our parallel universe just how many is anyone's guess. And that again is a whole new question. When we use the word 'aliens' we are talking about beings unlike us from other worlds. And that's the key word here 'other worlds' I believe there are thousands of other worlds living side by side in any number of dimensions. It is not just a case of us and an alien race it's case of us and maybe thousands of other life form. That would explain why people who claimed to have seen them have so many different descriptions. So you see if what people are saying is true then we already know there are many out there. Some may well be friendly and other not so. But I think if they do make contact then we should be very ready and hope that we can communicate and understand each other and help each other. Because if there is a 'war of worlds' we would lose hands down.

The ancient Egyptians

There has been a lot of talk about the ancient Egyptians and how the pyramids were built. This is something I also think about and for a long time I have believed it was the work of others. My own take on this is that beings from other worlds did indeed live here at one point. It may well be that 'Stonehenge' along with many other ancient sites were also built by such beings. The question is when? And were they here before us it is very possible that they were.

You can argue that these sites have been dated and we know that humans were here. But I also think as I have already said that our dating has been wrong many time also our understanding of the writings and art on some of these sites all seems a bit too clever. I don't believe we really understand them and again I think there is a lot more we simply don't know and may never know. It's almost a guessing game I feel we are not looking in the right place or the right way. Maybe we need to sit down and re-think what we think we know and try a new approach.

The thing about the ancient Egyptians and others like the Mayan and Inca and Aztec and probably many other peoples that we have not found yet is that they were very different to other people that we know about. One might even say that they were either put here or were a part of something much bigger. Certainly they had many rituals and customs relating to alien life and on the evidence it would seem they knew a lot about things that we are still learning about. The problem here is we humans like to think we are so clever but in truth we're not but we could be if we just hold on a bit and re-think.

In my first book 'life after death' I spoke about what happens when we die. And I said that we all meet again in another dimension as a mind as appose to a walking talking person. And we all will go to another place another dimension as all that have lived before have gone. You see like I said in book one the mind or soul whatever you call it has to go somewhere along with the minds of the billions who have died. So we then become something more than human we become another life form. Crazy but I believe true, so you see there has to be other life out there in as many dimensions and you can imagine.

What if a UFO lands on earth?

Depending where it lands my worry is that if one should land the first thing we do is surround it with guns and tanks. Not the best idea or even try to destroy it before we give them a chance to communicate. But let's assume we don't do the above then I think we must wait for them to make the first sign. I think that they may well understand our language or some sort of symbols that we both will understand. It's also important that we understand that for them to go between dimensions and have such crafts to do it in that they must be very advanced so we should not think we are the greater power. And again I have to say we must be ready as there are so many life forms we won't know if they are friendly or what so being ready is key here but also being ready to accept them in peace and find the common ground as to what they want. I do have to say here that I can't help thinking that if they do land it won't be for tea! And if they bother to come they must have a good reason. We could even find ourselves enslaved by them and they take over the earth. But then I also have one idea that I have had for many years. That they are watching us and visit from time to time waiting for the moment we are on the edge of destroying our world. And at that point they step in and stop us. I think this because I have an idea that we were put here as an experiment by them to see if we could evolve and create a society that could govern itself and exist in peace. The above is just something that I keep thinking about almost like we have been reincarnated from where we go when we die to this earth as a new race. It really makes sense maybe this is why I keep thinking about it. Ok let me explain what I mean as if you have not read my first book then you maybe are lost at this point. So when we die it is my belief that our mind/soul whatever you call it. Goes onto another dimension where we meet all those who have died before us. Ok so it maybe that whoever is in control in this dimension (the cosmos) somehow created the race we are now a part of a sort of mess reincarnation and as we know reincarnation does exist and here we are and the cosmos just keeps adding to our numbers. Almost like recycling! And so as I said these UFOs pop back from time to time to see how we are doing. Yes it sounds mad but don't dispel it just because it sounds mad. But it is all very possible. The point here is this, don't think we are the only race in the cosmos and be open because we don't know what is out there or how it works. I am giving my opinion and what I believe. And as crazy as it may sound it may well be true in part or even in fall.

Are there aliens living among us?

I know there are many UFO hunters who think there are aliens on earth with us. But as I said a few pagers back the form we are reflex's the environment we live in, so if they were here we would spot em! You know when men go to the moon they must live in a space suit or die. It's the same with aliens they come from an environment very different from ours so I don't think they could just walk around unaided and take a job in a fast food shop!

But then again you have to look at the possibility that they were human ok I joke. But really there is no reason why they should be here even if they wanted to be. I mean look around earth is not the best place to be right now. UFO hunters love to think up all these wondrous ideas. But really the best they do is speculate. If you watch any of the UFO shows on TV not once have they shown any real proof of anything. It may make interesting TV but nothing more. But I do like watching these strange people with their strange theory's and when they interview someone who claims to have witnessed something. I know I sound so negative but I'm not I know UFOs and other life form exist I just don't agree with these people. And as I said we are looking in the wrong place.

Do governments know anything?

The problem here is governments lie all the time who the hell knows what they know or don't know? But I don't think they have much to go on if they do indeed have a spaceship as many believe. Then that's all they have and nothing more. I do think they know something or have found some interesting things in space. But think about it if they were in contact with aliens then normal life here would change because I don't believe that after contact any government would be able to hide it. And also the state the world is in with wars everywhere. If aliens were here or governments knew anything then I think the world collectively would be standing together to make our world a better place. I mean aliens a higher power a bigger intelligents race here on earth it would be the single most amazing thing to happen and one that would untie all humans. We would have to stand together as one in light of a risk to our own world. But as you see 'nothing' we still kill and make new ways to kill nothing has changed. But one day I am sure it will be very different. If one day aliens do come then we would all be waiting for what comes next? Also if they came here ask yourself would you believe your government or would you now start to follow the new intelligents who may well offer a better existence.

There are so many 'ifs and buts' none of us knows enough. So let's move on from 'aliens' and look at why some people are born to be great thinkers and inventors and the like.

Reincarnation the great minds of our time

Have you ever wondered why you are not a great inventor or composer or scientist? Well here is what I believe. We are all born the same with 'if you like' the same chances in life. Some say that it depends where you live and who your parents are and rich or poor. In a word 'crap' I believe that the great minds of our world were all here before they lived and learned. Then when they died they were reincarnated to carry on their learning. And over many reincarnations they finally became the geniuses we know and love. I believe this because being born equal there is no way one person can know so much in such a short time. Because it takes a long time to be old enough to get to an age where you start to think in such a way that you become a great mind. Some of the great people we know are so clever that they must have been here time and time again. When you think of a 4year old who can play the entire works of Mozart or write a music score or play. It's crazy to think they were just lucky. But to say they were 'born clever' is closer to the truth because that's what it is they were born clever having been here so many times before to learn that 'yes' now we can say they were born clever. So you ask 'what's this got to do with other life forms? The fact that they have been to the other world where your mind/soul goes too. And there they have learned more because again if you read book one you would understand what I am saying here. These great minds have been on the other side and have learned a lot then have been reincarnated to continue and advance the human race. Because as I said in book one life is about learning and experience. And if you think about it one life is not enough time to know so much and yet these great minds do because they have been here many times. Let's take a life here ok and see how what I am saying works. A normal child.

A child is born up till about 4 they know nothing they are a kid. Between 5 and 7 they go to school learn the ABC maths ext. 10. 11. 12 13.14. Teenager doing what teenagers do. Then higher school say 5 years then collage and so on there is not enough time to be as incredibly clever as some are. Ok many people become doctor's ext but to be like an Einstein I do not believe you can learn all he knew in one life time. So it makes sense that we are reincarnated time and time again. And when a great mind dies he/she will be reincarnated again to carry on from where they left off. 'Not as mad as it sounds trust me'. You may ask why you or I are not a great mind. Well I got news for you! You may well be on your way to be such a person. But the problem with reincarnation is that you won't know you have been reincarnated. Ok I know there are a few cases where people have known things from pass lives that they couldn't know in this life. But in most cases you won't know. But in any case the thing is that the cosmos is made up of many life forms and we too when we die become another life form in another zone. Much of this may sound bizarre but only in as much as we really don't know and anyone who is thinking how stupid all this is. Should think again as much that we now know and could not believe are now fact. I'm not saying I am right but in truth no one can say I'm wrong.

New thinking

What if all I have said was true, what does it mean for us and the world we know? I can see a time coming when one day our world becomes a part of a whole new galaxy. That is to say that some where somehow other life forms will join our world and we and others from far off planets will share the galaxy. And all will travel between worlds. I also think this has already begun we are being watched and at some point contact will be made. It is all just a matter of time before our earth becomes home or at least a holiday resort for others.

Time travel

This is not something we need to look far for. As unknown to many or that we don't realize but we have been doing this for some time. In little steps things have moved in time and space. What do I mean? I'm sure we have all lost something and found it and stood for a moment thinking 'strange I'm sure I had it a moment ago or that you felt that things around you for an instant are not what they seem, it's like seeing the same person in the street 4 times in a few minutes and all around seems as it was the first time. What I'm saying is time and space do not work at the same speed and at any point it can and does change. Time bends it moves it plays tricks with us not knowingly but because we don't understand it. At anytime you may walk in and out of a small time gate. There are I believe thousands of these time gates but because we can't see them or even believe they exist we simply carry on and ignore the feeling.

In years to come we will find/discover a time gate somehow one will be found and then once we find one we will find more and more and bang! We will see the gateway to other worlds. How we deal with this is anyone's guess but be sure it will happen. 'Your holidays will never be the same again'

I believe as many do that we are living in a parallel universe with many other life forms right by our side. Find the gate and then say hello. Sounds easy and one day when we have found our first 'gate way' we will wonder how we didn't find it sooner 'all in its own time' so ok time travel what more can I tell you? Well as I said holidays won't be the same. We will visit these new worlds as they will visit ours. We will meet and who knows maybe even form relationships and next copulate and form new cross beings just like we do now with people from other cultures. All this I am sure will be in the future how far away depends on how quickly we find our first gate. My fear here is that so far we humans have not been so good in how we meet new life. For example Indians, tribes of the Amazon, we find and take all that's valuable. And I am sure these other worlds will have much that is valuable. We have all seen the films of what happens when we went to that planet and wanted to kill all because of what there planet had that we wanted. Question is will we ever learn to be a part of and not the ruler?

Throughout our history we have killed and taken over all that we discovered and killed all that were in our way. Well next time it may not be so easy or we may find that we are taken over. And that's why we need to learn because nothing lasts forever and our time as 'kings of all we find may well be over. And then we will be ready and the cosmos or some great mind will then discover the first time gate. And like I said as for when well that will depend on when the time is right. But be sure it will come maybe not next week but it will come.

I have to say here that I do feel all this is going on now in other zones. And we are the odd ones out maybe even the missing part of the world story. UFOs and there people watching us to see when we will be ready. Maybe its they who show us the way.

War and peace in the new galaxy

To start with there has always been war on earth and in all the years we have been here not a day of peace has been known. In the new worlds of the galaxy I believe they do not have wars because they have a system or understanding that there is no room for conflict. But if as I said there are unfriendly beings then who knows but I don't think we need to worry. I believe we need only to be sure we humans don't start trying to take over or start killing as that's all we seem to be good at.

Having said all that one has to realize the cosmos is so huge you cannot imagine as there is no beginning or end it just grows and grows. So you see there is enough room for all to live. And all the time more new life forms are added by us as we die by others as they too die in whatever form they are.

What will these new worlds look like?

Good question! Well look at our earth how beautiful it is the animal, plants, sky, nature the country side. Other worlds will be as different as you can imagine but equally as beautiful. I am sure we will stand in ore as we look out over space landscapes and the wonder of how fantastic the creation of life and worlds really is. Just as they will look in wonder at our earth (I think they have seen our earth already) imagine standing on another world and looking back at earth or to the next world. We can only guess at what it will be like. And think also at meeting other beings. Think of all they do how they live what do they do for fun? How do they teach each other do they have schools? What do they eat how do they cook what names do they have, do they know 'love' what can we learn from each other? Reading all I have said and you can't help but see that we are very much the baby in the great plan of life. We have not even started to walk yet. I am sure the great minds that have lived and that will live. Will open our minds up to a million things.

Illnesses and cure

On our earth we have found many plants that can cure many ills. But still people die of many more things. What about in these new worlds they may have such advanced medicines that all human ill's can be cured. And just maybe we can help them. It's all about learning and sharing and in the end being one big family. What about the idea that in the new worlds people who are ill simply step into a box and are treated and cured in moments. No way I hear you shout! 300 years ago an x-ray was also a dream unknown. And again like I said if they can move between dimension to dimension then they probably have things so advanced that we can't even imagine. So you see when all this happens we will truly be one world helping and learning together. All ill's gone but maybe new ill's to come but together we can build a world that thinks of life and not money a world where we matter not material things a world where we use what we have to better us all and keep life alive. Just look at the counties on earth who kill their own with no regard for life is this really all we are on earth?

Don't you ever dream of a better world? Well I believe there is a much better way. And in some years to come so will you if we live that long to see it. No one can say that there is not something more something better out there. Because there is and it's waiting for the moment when we to become a part of it.

How will we get from one world to another?

It will be as simple as walking through a door. We will develop new machines a cross maybe between a bus and a plan 'airbus' ok I don't know but we will develop such crafts that can move between time gates. It's not so hard as we have the knowhow we just need to improve on what we have. So it works like this... you get on the 'star ship' it then takes off and heads to zone

4 time gate 4. As it goes through we just sit back and enjoy the view and ride. When we get to zone 4 the 'star bus' lands next to its landing station we get off and bingo we go to where ever it is we are going. Hotel ext.

Ok a moment as by now you are thinking either I'm truly out of my box or that 'hay' he may have something here. All I say here is what I think is to come in many years time. Don't think that all you see around you is all there is. That would be so stupid because there is a whole new world and worlds out there waiting and we humans will be a part of it one day.

A world like ours

Something that interests me more now than it did as a kid. That is years ago when I was 13 I dreamed of a world the same as ours only without all the crap. It was more than a dream as I even had an idea of how it worked and even where it was. Now I am 53 years old and a few months ago I was watching a program and the scientists were describing the very world I had dreamed of all those years ago so I will tell you about this world. So it was that I dream that in our world we have the earth then sky then space then deep space and some way further we have the reverse that is to say another deep space then space then sky and then another planet. This planet was the same as earth with people the same as us. What I mean is an identical me and you but on this planet there was no ill's not war nothing they live in total peace. And it was that they put us here on earth (I said something like this at the beginning) to see if we could manage ourselves with law and order. But we can't we failed and that is why UFOs are visiting us watching and waiting for the point where we are just about to destroy our world. And then they will make contact and show us that it was they who put us here all those thousands of years ago. And in telling us they will also tell us that now they must take back the control and make our earth the place it was meant to be. Ok so that was my dream but strange how now 40 years later others are thinking and talking about the same. Maybe not all but as far as this other planet is concerned they are talking about it. But it is all so possible as I keep saying just because it's not here yet don't think it won't be.

How will we communicate with other life forms?

This I think is the least of our problems as already we have ways of talking to each other without words. We use sign language as well as art and body language and music. And let's not assume that those of other worlds have not evolved much the same. The question is more about the way we communicate meaning we don't want to show we are unfriendly or that we show a position of power. We must show we are open to the new and ready to be a part of what is going on. We want these 'aliens' (for want of a better word) to see us as just a new form and not as a fret. Of cores it would be silly to not be ready for whatever may come but we can do this without having the army and guns pointing at their heads. It's easy to say all this here but if and when this happens we also have to deal with the unknown and our own fears. I mean if a bloody great space ship lands at the back of a school or even many land then it will put the fear up us all. But if we have a plan of action a way of being ready then it will go a long way to make it a bit better. But they too will have a degree of fear (I hope) so we will all be facing the same deal and we must on both sides be ready to talk and understand and find the common ground. Being ready does not mean blowing the crap out of anyone it just means be ready for all that could happen or that could go wrong. In the end whether we like it or not I believe it is just a matter of time and we can't avoid it will happen. There is no running away from it we have to meet it head on and deal with it.

Space pets

Strange don't you think that we tend to think of aliens or other form as so different from us. So as to 'do they keep pets' I'm sure they do and again these pets will look a bit odd. When compared to our dogs and cats. But then again they may have their own type of dog/cat. I am sure many of the new life form out there will live a bit like us so taking Fido for a walk will be the norm. It will be very interesting to see what sort of animals they have. Another question is do they have zoo's? What do they do for entertainment? What about music? And are they really going to be so different from us? You see when you look at it like that you can begin to see that maybe we have more in common than most think. It may well be that they work have jobs and kids. They get up at 8 and get the kids off to school dad goes to work mum goes shopping. At 6 they are all home for tea and talk around the table. Kids go out meet friend go to parties and come home late only to be told off by dad. Weekends they have a BBQ drink space beer and fall over. It's easy to imagine they are just like us. And as I said there are so many new worlds out there some are bound to be like ours.

How do the move about in there world?

Here on earth we have cars, bikes, scooters, trains, bus, ext. I am sure they have much the same but not as we know it. I mean they may have hover bikes and cars or even simply just think themselves to where they want to be. There will be as many ways as there is worlds. Each new world will have its own system of getting from a to b. The thing is here is that we watch space films and so most think that's how it will be. I really don't see any reason why they should be any different from us. Apart from how they look as their shape will depend on the environment in which they live in. But apart from that I am sure they have fun are happy love what they do as well as hate working paying tax and complain about who is running the planet. And if you think all this sounds really stupid then look at us! Maybe they are looking down on us thinking 'different world same shit' now you can see in reading this just how much we don't know about other worlds out there. All we see are stars no way can anyone say 'your book is crap' because no one has any idea what is out there. And it just may be I am right. So let's just take a quick look at what we do know about life and other worlds. We know that spaceships are for real; **there ya go that was quick!** *The thing is we are guessing at what may be out there but we do not as yet have the technology to see a thing. In all the hundreds of years we have been looking at the stars we still know nothing. But we guess a lot at what maybe out there. Every TV show you watch on the subject of space is fall of 'ifs and buts. Theory is a good thing as is 'idea' I mean many medical breakthroughs were discovered from a single idea or theory so it's not a bad thing at all. And indeed this book is a part of the very same idea. But the bottom line is 'we just don't know' and until another world makes contact with us or we find the first time gate we will just keep guessing and looking in hope that something turns up.*

The hunt to prove they exist

There are many clubs and individuals who spend their free time looking 'up' and sometimes 'down' or even 'around' all with a take on how and when and where. UFO hunters (don't ya just love em?) if you've seen the TV shows and I am sure you have. They always go back to the same story and places. Hoping that somehow something new will be found. Some of their ideas are as bizarre as they are. (Sorry) I guess if there were not doing this they would be down the pub playing darts. It may make good TV (I don't agree) but really it is a huge waste of time. They talk to people who have a story that turns out to be crap. Or they walk over the same old ground with a new machine and 'nothing' a few pictures and a lot of 'ifs and buts' the only thing we can say that is fact. And we really didn't need to research it because there have been cave drawings and sittings for 1000s of years. UFO are real and therefore so are other life forms as someone must be flying or controlling the craft.

UFO sightings

I think most of us have seen something up in the sky that we couldn't put into any group of flying craft that we know today. I myself have seen a few over the years. I can't tell you what they were or what it was only in as much as I had not seen it before. A trick of the light? Electro field? Clouds? Reflection? Eye's playing tricks on you? At what point do you say 'A UFO' many sightings and pictures are to be honest so far away it can be anything. We live in a world that is fall of optical illusions we are always seeing things that are not there or are not what they seem to be. It's a human thing but then comes the sightings that we can't ignore. I guess I have seen about 20 things in the sky over my life. But out of the 20 ish I can say that only 3 made me really think and start to look into UFOs. My first real sighting was on my way home one night and this light I am sure was following me (or seem to be) I watched it for about 10 min before it went. But more recently in 2012 my girlfriend and I were coming back for the dentist walking. She looked up at what was like a fire flying across the sky changing colour as it went. It was in control that is to say it knew what it was doing. It didn't burn out so I knew it was not a comet. And when looking at the back it seems to be burning gases. It could have been any number of things. But something inside me said 'UFO' was it? Another sighting I had was at the age of 17. I was walking over a bridge near where I lived. I stopped to admire the view of the country side. And there it was as clear as day so clear my girlfriend and I started to walk back at where it was in the hope of seeing if there was anything more to see like life forms. That's how real it was there was no getting away from it what we were looking at was a UFO and it was beautiful the colours and it seem to be saying 'peace' we both just felt at peace no fear at all we were actually running towards it. We got about 200 yards away and it was gone just went not up or down but like it was never there. We question each other about what we had seen and just stood there I think she was crying not from fear but the beauty. So I think this is the reason I am writing this book as I believe that there are other worlds and that many of them will be so friendly and just maybe the reason there not with us yet is their own fear. I mean once they are met and seen there is no hiding anymore. And trust me who the hell wants to know the human race? Just think for a moment you are looking down on the human race. What can you see? A beautiful earth with all the colours of a beautiful nature, then you see humans killing each other in as many ways as you can think of. Death and destruction the fall horror of a world consumed by greed and power, where life means nothing.

Alien conversation..... *Em nice planet let's not take the kids there dear let's head too 'zogg' zone 7 we had a great time there last year. Oh dear do you think they saw us? No love and if they did they will think it was a weather balloon! They don't look so clever or advanced I mean look how the live and kill each other so old hat no new thinking there 'I thinks' conversation ends....*

I think it helps to look at both sides of the story, why do we fear aliens? They just may have already made their minds up about us. I think if I were them I wouldn't want to come here either. So maybe they fly past us now and then just to see if we are still doing the same shit to our planet and to each other. No points for guessing the answer! So you can see there are any reasons why contact has not yet been made. So we go back to us finding a time gate which may be the only way we move on in our quest for other life forms.

Why believe we are not alone?

When you think about it there are many life forms on earth and even more that are yet undiscovered. In the oceans and forests in the earth itself. So it's not so silly to think there are beings out there in space. The cosmos is infinitive you cannot begin to imagine its size in a word 'it just is' so if you think that we are all that exists in this huge universe then you need to wakeup. There are billions and billions of stars and planets and beyond them there are even more and beyond ext. And as sure as eggs is eggs somewhere in among them are beings living lives. Doing whatever they do but they exist and are as real as we are. We humans believe in many things that have little or no real prove. But UFOs we know are there as they have been seen by many 'many times. It is probably the one last true wonder yet to be found as we know it's there we just don't know how to get there. I am sure they know about us as they must do as we have seen their crafts popping in and out of our time zone. Why they don't make contact is anyone's guess maybe they are in between time zones and don't see us who knows. But for sure we are aware of each other's existents.

To end this little look at 'other life forms'

Remember for as many years as we can say UFOs have been reported. From as far back as the Egyptians and others like the Mayan and Inca and Aztec UFOs have been noted. We in the 20th century are just another age that has seen the evidence. Weather you believe or not one fact is clear we are not alone never have been. And for now all we can do is wait and keep looking for those time gates. I hope they or whoever finds them soon so I to can look with wonder at other worlds and other life forms. One day I am sure the world will open its secrets to us and we will become a part of the whole cosmos and welcome the dawn of a new age in the history of man and the world we live in.

Printed in Great Britain
by Amazon

79812045R00031